One Cure:

An Essay Offering an Alternative to the View that a Single Cure for Cancer is Impossible

By John R. Freeman II

Contents

Introduction

I do not believe that scientists and medical professionals are gods. I do not even assume that they are right about a lot of the things that they hold for granted. I feel that in the future, say, a few hundred years from now, our descendants will look back on our medical understanding of the world in the same manner that we look back on the charlatans of the past.

Imagine your stereotypical image of a salesman in a horse-drawn wagon in the West selling elixirs to the unwary. That's how I view our current understanding of medicine. On a sidebar here, that view applies to all of science: in a recent survey it was discovered that over half of physicians, dieticians, and physical trainers (who would be expected to hold a scientifically-accurate view of at least weight control) thought that when you burned fat, the fat was transformed into heat and left the body that way. Never mind that that violates the conservation of mass – the answer, of course, is that the fat is combined with oxygen to produce carbon dioxide and water – you basically breathe and sweat it out. It's simply mind-numbingly insane that a majority of people that you would go to and expect sound advice from on the subject of weight control hold a false view of how it works.

My view (of science in general – not weight control – obviously) is based, in large part, on the book *The Structure of Scientific Revolutions* by Thomas S. Kuhn. It's a fascinating book. To simplify it: mainstream science is simply the view that the majority of scientists in a field hold. That view is the paradigm within which they operate. True advances in knowledge are usually made by individuals working outside of their field or that are new to the field and are made by accident. That is, the great thinkers that actually brought us closer to a more complete understanding of the world were doing their work as a hobby or hadn't become dependent upon defending an old point of view to make their living. They were not part of the system. I feel like a lot of what we think about cancer and various other diseases is like that. The system is preventing the next advance, the next great leap forward.

The other part forming my viewpoint comes from Anthony M. Alioto, a professor at the college that I attended. I forget the name of the class (something like A History of Western Science and Religion), but he taught from Kuhn's book as well as from his own, *A History of Western Science.* His book took a lot of Kuhn's ideas and expanded upon them. It's a pretty good read.

I would like to start by saying that I am not purporting to have a cure for cancer. What I am writing about in this essay is the route by which I believe one can be found, although I do so by describing a treatment that I call reverse somatic cell nuclear transfer, and then discussing the ways in which it can be accomplished, the hurdles to doing so, and the possibilities that open up if it is possible.

This is not an academic work. There's a reason for that. As far as I have been able to tell, my idea of reverse somatic cell nuclear transfer is original thought. Granted, it's such a simple idea that someone, somewhere, and at some time must have thought of it as well, but if they did I have not been able to find evidence of it. You will only find a citation(s) where I am borrowing someone else's idea – usually so that I can talk about what I do not like about it. Then, even when I credit someone else's idea I don't bother to follow any accepted manual of style.

Anything that I give credit to can be found with a few minutes of your time and Google.

I put a lot of effort into maintaining a conversational tone. There are a lot of 'fucks' sprinkled throughout. There is also a lot of sarcasm. That is how I communicate. If you are looking for a thesis, for a general idea behind the entire essay, then I will give it to you now:

Too often we confine our thinking within boxes. However it is important to realize that those boxes do not exist except within our own minds. If you think that something has to be done a certain way, then you are limiting yourself to only doing it that way. You've confined yourself within a box without realizing it. Only once we have learned to ignore the boxes that are impeding our thinking can we truly begin to imagine a better world.

On that note, I hope you like the ideas that I am putting forward. Hopefully enough to share them. Or not, in which case fuck you too. I plan to simply self-publish this essay without any marketing and see if anyone reads it. I assume that if you find the idea compelling that you will share it with a friend. And then they'll share it. And so on, until someone actually in a position to do something with it reads it and changes the world. Or fails, but then at least they would be doing something.

A Lack of

Imagination

"No such thing"

"This is no such thing as a 'cure for cancer' because…"
Any quick search for a cure for cancer will end with a long
list of supposed reasons. Well, unless it's by a person
peddling the snake oil known as supplements or 'natural'
cures. They'll be glad to take your money in exchange for
something that will never work. My own grandfather was
convinced that all he needed to cure his colon cancer was
goat's milk. Can you guess what they cut out of him a few
years later?

Oh sure, they (they being the real scientists – not the
snake oil peddlers) have convincing arguments for why
there can be no such thing as a cure for cancer. They
dwell on how every cancer is different, or how different
drugs work differently on two people with the same type
of cancer. It isn't so much a disease, they say, as it is a

collection of diseases lumped together within the same category.

One short story from my own life will highlight what is wrong with this kind of thinking (on a side note here, one of my friends that reviewed this essay for me recommended that I not put the following story in because including my political views would turn people off of actually getting around to reading about my idea for a cure. It remains included because the central idea, that prevention is not a cure, is important to my argument that the cancer research field is stuck in a mindset that doesn't truly believe that a cure is even possible). During my junior year of college the extracurricular group that I was part of forced me to take part in the American Cancer Society's Relay for Life event. I had to 'donate' $10 and participate in the walk. I thought the whole thing was stupid, but paid my fee and took part. During the application process they asked for my email address. Months later I received an email from them asking for more money. It's important to note that I went to college in Missouri. Missouri is notable because they have the lowest taxes on cigarettes in the entire United States. At the time they had a ballot initiative coming up for voting that would raise the tax on tobacco to some outrageous amount. It was all over the radio. Every time that I went into the local smoke shops I would see posters decrying the huge increase. Since they were claiming that the extra money raised would be used to fund schools I fully expected the measure to pass.

Let me be clear: as a heavy smoker, libertarian, and American that actually believes in freedom, I think that people that want to use the government to punish my (or anyone else's) personal choices are in the wrong and can go fuck themselves. It's an unfortunate side effect of living in a democracy that no one really respects individual choices, and if they can use their distaste for my activities to steal my money in the form of taxation they'll feel even less respect for me. I've made peace with the fact that my fellow man are tyrants and deserving of the eventual and much needed revolution. Hopefully those that believe that compulsion is acceptable so long as enough people agree with you will be the first up against the wall when it comes – right beside those that agreed.

So anyway, during all of this I got an email from the American Cancer Society wanting more money. Now, if you look on the Relay for Life website they go into how your money is used for research, and to support various survivor and recovery programs. I had read all of that when I gave them my $10. I mean, who doesn't want to support cancer research? However, in the email that they sent me a big deal was made about how they had helped fund the initiative to put higher taxes on tobacco on the ballot in Missouri. Hmm, I thought, what the fuck does that have to do with cancer research? Needless to say, they did not get my money, nor will they ever get another penny from me. I will say this for Missouri: they defeated the measure. Good for them. My own home state of Texas took the easy money instead of standing up for Liberty.

What cancer is

To understand why everyone who seemingly matters (the 'experts') believes that there will never be such a thing as a single cure for cancer, or why they believe that cancer is a collection of hundreds or thousands of diseases (or even possibly a disease unique to each person that develops it) that must be defeated individually, it is important to understand exactly what cancer is. Cancer, at its most basic, is an uncontrolled growth of cells in the body caused by mutations in the affected individual's DNA (that angry gasp you heard came from all the cancer epigeneticists reading this – I'll touch on epigenetics later, but for now just assume that I'm talking about mutations in both the nucleotide and in the person's epigenetic structure). The article "The Hallmarks of Cancer" by cancer researchers Douglas Hanahan and Robert Weinberg lists six characteristics of cancer.

The Hallmarks of Cancer

The first characteristic is self-sufficiency in growth signals. Normally a cell requires a reason to divide. These reasons are given to it by external signals called growth factors. Lose a chunk of skin in a wicked motorcycle accident? Growth factors signal to the surrounding tissue to grow and divide, filling in the lost area. Hopefully. In a cancer cell, however, the cells grow and divide without any external signals. Their division is stuck at 'ON'. This is why cancer cells continue to make more cancer cells that form tumors and begin to crowd out neighboring cells.

The second characteristic is insensitivity to anti-growth signals. Normally, when your cells have filled in that chunk of skin you lost and the cells sense the presence of other cells surrounding them, they respond to signals that tell them to stop dividing, which is why you don't develop a tumor every time that you get a paper cut. In a cancer cell, however, this signal is ignored and the cells continue to divide.

The third characteristic is that it evades apoptosis. Apoptosis is programmed cellular death. In normal cells, once its DNA becomes too damaged the cell will destroy itself, preventing the damage from becoming the new normal in its descendent cells. However, cancer cells manage to evade their own mechanisms for detecting damage, like having a watch dog with a selective sense of danger; it didn't bark while your house was being robbed – it thought the robbers were its best friends.

The fourth characteristic is that it has limitless replicative potential. Normal cells are governed by the

Hayflick limit. That is, that they can only divide a certain number of times before they enter senescence. That number is usually between forty and sixty times and is governed by telomere length. Each time a normal cell divides, its daughters have slightly shortened telomeres. When the telomeres are too short, the cell no longer divides. Cancer cells ignore this natural limit on replication and spawn an infinite number of generations.

The fifth characteristic is that they possess angiogenesis, which is the ability to make new blood vessels form, feeding their growth. I won't say too much about this one since cancer cells initially lack this ability and I don't agree that angiogenesis is a characteristic of cancer. Do cancers display angiogenesis? Yes. Is it a function of something that the cancer cell actually does, or is it a function of simply having a new mass of cells grow? I don't know. If you suddenly grew a third kidney it would presumably grow an accompanying blood supply, but it wouldn't be cancer.

The final characteristic that they outline is that cancer cells have the ability to invade other tissues and metastasize. That's the one characteristic that really makes them dangerous. Have a tumor that's growing? If it can be cut out then you're fine. Suddenly have tumors forming all over the body because the cancer has reached metastasis and pioneer cells are using your blood as a highway? You're pretty much fucked at that point.

A more simple definition

Now those are some good characteristics to attach to a definition of cancer, but when you get right down to it I believe that the definition of cancer can be narrowed down to a smaller number of characteristics. The first is that mitosis is uncontrolled. Cancer cells divide and continue to divide without regard for any calls from neighboring cells or the body. Instead of being, say, skin cells that get along with their neighbors and stay in one place replacing their natural losses, they multiply and burst out of their points of origin, forming tumors and crowding out their neighbors. The second is that the cells have gained immortality. Cells have a limited number of times that they can divide normally. If the cells that are rapidly multiplying do not develop immortality, then they will quickly reach the end of their reproductive limit and the tumor will stop growing. It will be benign. The third characteristic is that cancer is localized. Cancer cells, even when they become malignant and begin to colonize the rest of the body, are unable to turn other cells cancerous. The fourth and final characteristic is that the uncontrolled mitosis and immortality are caused by genetic mutations.

Perhaps, when you get right down to it, the single defining characteristic of cancer is that it is caused by mutations that a person does not possess at conception, but instead acquires throughout life. Granted, it is entirely possible for a zygote to possess mutations that will cause cancer, but they quickly render the fetus unviable. Any zygote formed from the combination of its parents' DNA that is viable will, by definition, be cancer free. We all start out with non-cancerous DNA and, indeed, the vast majority of our DNA remains non-cancerous.

Why a single cure is believed impossible

There are some genes that, if possessed, raise a person's lifetime risk of cancer greatly. Normally, it takes several mutations in a person's DNA to occur before they develop cancer. If they are born with a cancer gene or genes, then they start out in life with a handicap. Fewer mutations – if in the right locations – will have to occur in their DNA before they develop cancer.

These cancer genes are the focus of a lot of ongoing research due to the fact that they are seen as a means of combating certain cancers with known causes. For instance, if you can create a drug that specifically reactivates a gene that is inactive due to a known mutation or that counteracts the effects of a mutated gene, then you have created a system of managed care with a captured market. That is why there is so much emphasis on testing for known cancer genes such as mutated *BRCA1*, which raises a person's risk for developing breast cancer greatly, made famous by Angelina Jolie's decision to undergo a mastectomy due to her testing positive for a defective *BRCA1*. Given that people with a defective *BRCA1* have a 75% chance of developing breast or ovarian cancer in their lifetime, if someone were to develop a drug that would target the defective gene and make it operate normally, then they would suddenly have a market of *all* the people that test positive for it. Hell, just the test itself is lucrative, with the testing service being sold for $4,000.

That said, cancer genes account for only 5 to 10% of all cancers, and are possessed by less than a single percent of the population. The rest are considered to be environmental cancers. Since environmental cancers are not tied to any one specific gene, but are caused by several mutations across any of the genes in the cell, the consensus among scientists appears to be 'fuck it, stop doing what you're doing, we'll gradually work on it as we stumble across solutions – in the meantime, advocate prevention by avoidance'. Personalized medicine is the closest that they seem to come to trying to develop any cures. Since not everyone with, say, environmental pancreatic cancer has the same mutations in their DNA, then some of them will respond to drugs that target their mutations, while the rest will not respond at all.

Where they have seen results is with prevention – or rather avoidance (or as I call it: not really living). Somewhere above 30% of all cancer deaths are considered preventable so long as people avoid their causes. For instance, 90% of all lung cancer deaths are attributable to smoking, which is why so much effort is poured into trying to socially convince people to give up the habit. Where that fails, they use the legal system to compel people, with smoking bans inside of 'public' places such as restaurants (or places where it makes no sense to be if you're not smoking, such as a bar – smoke in the air adds atmosphere; without it you're just in a building with a bunch of drunks), outright prohibition in certain areas and for the young, and the ever-popular increasing of taxation on tobacco.

As I've alluded to before, I don't believe that it is the purpose of science to be used to *compel* behavior. It can be used as part of an individual choice to *influence* behavior, but compulsion is morally wrong. I like to think that we as a society have embraced the idea that might does not make right. We should also embrace the idea that *right does not justify might*. You can be right all fucking day long about my habits and it should never justify you interfering with the personal choices that I make with my body. The right to do what I want with my body should not apply only if I'm a woman with a fetus growing inside of me.

That said, the other known preventable environmental factors are obesity, poor diet, physical inactivity, sexually transmitted disease, and pollution. Holy shit, they really have opened up Pandora's Box by justifying increased taxation on behavior that increases cancer risk. Just being an American could potentially put you in the same boat. Am I saying that as a group we're a bunch of fat, fast food eating, city-dwelling slobs that get winded when we go up a flight of stairs and don't have a clue how to use condoms? Yes, that is exactly what I'm saying about the group. At least I'm athletic and thin.

Then there are the environmental factors that cannot be controlled, such as background radiation. Hell, they've proven that exposure to *oxygen* causes cancer. Good luck avoiding that one. Even more irritatingly, there are all the things surround us that we don't even know cause cancer. Hell, asbestos was considered a miracle material when it was developed, which is why it was in so many things. Hilariously, a cigarette company tried to reduce the dangers of smoking – by developing an asbestos filter. I'm assuming that eventually they'll prove that *water* causes cancer.

Regimented thinking and perverse incentives

The realities of the cancer research world have created an environment where researchers are unable to develop a cure for cancer. There's simply too little outside-the-box thinking. Maybe I'm being too harsh. Angiogenesis inhibitors were at one time thought to be a potential cure for cancer, and God bless them, they actually tried to use them that way. As I mentioned before, angiogenesis is/was thought to be a characteristic of cancer. If you can inhibit that characteristic, the thinking goes, then you can stop the growth of cancer. That was not found to be the case in practice. In certain situations it has been found to be effective in stopping the growth of certain cancers and when used in conjunction with other therapies, but it isn't a cure. And then there's the fact that angiogenesis is, well, *necessary* for proper bodily function. At least they were on the right track: identifying some commonality of all cancers and then try attacking it. It just wasn't the right one. Or possibly not even a commonality.

Hyper competition for the limited resources given to cancer research is the main reason why researchers suffer from a lack of imagination. Since the War on Cancer was declared in 1971, over $200 billion has been spent on cancer research for an overall reduction of 5% in the cancer death rate from levels in 1950 (personally, I am curious as to what the monetary cost of the War on Cancer would balloon to if you included the cost of prevention campaigns, enforcement, and the economic impact of lost business revenue). Original thinking is risky and expensive. It's a much safer bet for research money to be spent on small, incremental research that builds upon things already known. That's why you see bullshit about how 'research shows random product X that we all use increases risk of cancer!" stories in the news. Ok, it's informative, but it isn't new or original thinking. It's wasting money on things that common sense could have told you. Even worse, these studies are quite often used in calls for increased regulation or bans on products that – if we're honest with ourselves – are economically better to risk a slightly higher cancer incidence on than pay more for whatever 'safer' bullshit that's being peddled. Re:

asbestos filters. Seriously, there's no science quite like pseudoscience.

So, we find ourselves in a situation where unfathomable amounts of money are being spent on unimaginative science – the worst kind – while its findings are being used to compel behavior. Does that mean that there truly is 'no such thing as a cure for cancer'? I would say no. Cancer can be cured – as an entire category of disease – they simply have to break out of their comfort zones and ask themselves "what else does all cancer have in common and how can we attack that?"

A Cure for Cancer

The idea

So then, what is the weakness that all cancers possess? The answer is genetic mutation. Or, rather, defective DNA. All cancers (and tumors) have that in common: genetic mutations in a person's DNA have rendered it malicious, turning healthy cells into self-propagating monsters. If you can correct or prevent the mutations, then you can cure cancer.

My idea is that means can be developed to extract perfectly functioning DNA from the healthy cells in a person's body, replicate it, remove the cancerous DNA, and replace it with the normal variety. Of course, targeting *just* the cancerous cells would be problematic, so instead *all* of the body's cells would have their DNA replaced. A sort of total body genetic overhaul. Without their defective DNA, the cancer cells will cease to exhibit their abnormal characteristics: they won't grow out of control, they will exhibit apoptosis, and they will respond to anti-growth signals (the tumors would still remain, although I am curious as to what would happen to them – I suspect that they would act like benign tumors and have to be cut out if debilitating, but it is possible that the body might reabsorb them). The DNA is the key. Simple, isn't it? I call it reverse somatic cell nuclear transfer or RSCNT.

For a bit of background, somatic cell nuclear transfer is the process in which the nucleus of an animal's somatic cell is implanted into an egg cell that has had its nucleus removed. The egg cell then divides and multiplies with the donor DNA controlling it. Cloning, in other words. Dolly the sheep was formed this way. With my proposed *reverse* somatic cell nuclear transfer, however, you take the genetic material from the donor cell, multiply it, then replace the genetic material of the donor's cells with it. You're cloning the genetic material but leaving the body as a frame. Reverse somatic cell genomic transfer might be a more accurate name for it – hell, an even more accurate name for it might be reverse cloning, but that's irrelevant. What's important is the idea.

The experiment

I propose the following experiment that can be performed to demonstrate the validity of the idea: take a person with cancer and harvest some of their healthy cells and some of their cancer cells. Using mechanical means, such as are used during *in vitro* fertilization (a needle – although we are actually talking about intracytoplasmic sperm injection – which is what people think about when they think of *in vitro* fertilization anyway), extract the DNA from some of the cancer cells. Then, extract the DNA from some of the healthy cells. Inject the healthy DNA into the now-empty cancer cells. Now, you have two control groups – the unaltered healthy cells and the unaltered cancer cells – and one experimental group – the cancer cells with normal DNA in them. If my idea has merit, then you would expect that the experimental group would begin to behave like the healthy control group. Granted, it should be expected that much of the experimental group will be too severely damaged by the process and will die, just like some egg cells do during SCNT.

So then, if proven viable, how would you make reverse somatic cell nuclear transfer practical? The way that I see it, there are two general methods that can be used. They are genetically modified organisms, or GMOs, and nanorobotics.

GMOs

Genetically modified organisms are the technology that is closest to being practical as a method for RSCNT. We already use GMOs to modify targeted portions of DNA in organisms for the purpose of changing them - whether it is to cure a genetic defect, insert a desired gene, or create a new species. What I'm saying is that we have experience with modifying organisms. In other words, we have the technology, we can rebuild ourselves.

Viruses, of course, are the organisms that we have the most experience modifying to affect our desired means. If we are already able to use viruses to insert or delete sections of DNA or genes, then it is a logical step to use them to insert entire chromosomes. One step up from that is entire genomes.

The problem then becomes how to use them to remove the chromosomes that are there already. That's a matter for the genetic engineers to figure out. If I had to wager a guess, then I would say that at least two different organisms would have to be modified: one to remove DNA and one to insert it.

The technical questions then revolve around how to manage the process so that cells that have their DNA removed are quickly filled with replacement DNA, how to ensure that cells that have had their DNA replaced do not then have it removed, and then how to ensure that the GMOs do not become a communicable disease; spreading from person to person until we all have the same DNA.

Modifying the viruses so that they have the ability to chemically mark the individual cells is one solution to the first two problems. Perhaps they can leave a slowly degrading chemical behind, which we will call chemical A, for the DNA removal viruses. The DNA replacement viruses would be modified to only insert new DNA into cells that they detect chemical A in. After they are done, they would leave behind what we will call chemical B. DNA removal viruses would be modified to ensure that cells containing chemical B would be left alone.

The question of how to prevent person to person spread could rely upon another existing technology: unnatural base pairs. All DNA is made up of two pairs of nucleotides: adenine and thymine, and guanine and cytosine. Scientists have managed to develop artificial base pairs that are not found in nature.

I suspect that if you modify a virus to have artificial nucleotides — nucleotides that are created using chemicals not usually found in the body — in its genetic make-up, that you can prevent its spread. For instance, modify the virus so that it cannot function without creating the artificial nucleotides. It will not multiply, insert chromosomes, or remove them without collecting the required chemical ingredients to create artificial nucleotides first. Then, what you can do is infuse the patient with the chemicals before their RSCNT procedure. The patient has the required chemicals spread throughout their body. The viruses are injected into them, they use the chemicals to do their functions, and, when done, they are not able to spread to another person in the patient's life because that person does not have the required chemicals in their body. An even better solution would be if the viruses are created in such a way that they have to create unnatural base pairs in order to survive. Stop infusing the patient with the required chemicals and the viruses inactivate once they run out.

The exact method in which organisms can be modified to conduct RSCNT will, of course, depend upon what is actually possible. It might be that viruses are not able to hold an entire set of human DNA. How would that be overcome? Let's say that each virus can only carry one pair of chromosomes. Then you require twenty-three different viruses to remove DNA from each cell and twenty-three to insert DNA into each one. Give each one a different marker to react to and leave behind and shit becomes vastly more complicated.

An alternative to viruses could be to use bacteria instead. They are bigger, which is both a positive and a negative. A positive because you would have more space to deal with for the storage of DNA, but a negative because with their increased size crossing the cell membrane becomes a problem. But those are just thoughts of things to be overcome if bacteria are seen as a more viable alternative.

Nanorobotics

Altering biological functions, though, is messy, which is why machines are an eventual better choice for RSCNT. They can be programmed to fulfill an exact function. They do not mutate or evolve. They have an advantage over organisms in that they can potentially be collected and reused once their task is accomplished. The only problem is that the technology has yet to match its promise – it doesn't really exist as such just yet.

Eventually, though, RSCNT could be accomplished much more reliably using nanorobotics. Instead of having to muck about with modifying viruses to do what we want, tiny machines could be programmed to do exactly what we want, kind of like how instead of having to deal with strikes and lazy union workers manufacturers purchase robots. This also overcomes the problem of how to convince the viruses to collect their healthy DNA cargo in the first place. You would simply tell the machines to do so, perhaps by programming them to act in a certain way when exposed to a particular frequency of radio waves.

Another advantage over viruses is that they can be programmed to self-destruct, either after accomplishing their task or after receiving a signal, ensuring that the machines will not maliciously get out of control; recreating Skynet on the cellular level. In all likelihood, though, it would be better to collect the machines by having them exit the body through the urine when their mission is accomplished so that they can be reused, keeping costs down.

A different kind of cure

Another method of eliminating cancer (not curing it in this case – rather, we are talking about prevention – of a type that does not involve avoidance) would be to design and create a new organelle for the human cell (we're going well outside of current technological ability here – almost transhumanism). Make it a symbiotic organism with its own DNA, like mitochondria, and it gains the ability to be an inherited characteristic of the cell. Otherwise, we would have to dabble with genetic engineering to convince the human genome to do the same thing. This hypothetical organelle would need to possess an important and unheard of ability: the ability to pass between cells. My idea is thus: the organelle would be designed to travel between neighboring cells, comparing the DNA of them against each other. Since a genetic mutation on any one particular gene would be expected to occur within only one cell among a group of neighbors at a time (assuming that the organelles are introduced into a healthy, non-cancerous organism), the organelle would be able to determine which of the cells in any group had undergone genetic damage. Then, and this is critical, the organelle would be designed to either direct the DNA repair process or repair the damage on its own.

The upside of this idea is that it would add another layer to the DNA repair process. Because of the limitations in the DNA repair process, everyone – if they live long enough – will eventually develop cancer. If you can add another layer to the process, you would expect to decrease the overall chance of developing cancer in a person's lifetime. Make that layer a comparative layer, and I would expect it to decrease even more dramatically.

Here, too, we can use nanorobotics in lieu of organisms to accomplish the same thing. Instead of creating an organelle, we could instead design nanobots to do the same thing.   Granted, they would not be inheritable, but unlike deliberate genetic engineering on the scale required to design and create a new organelle, the idea is entirely plausible – if not entirely possible at this time.

Hurdles

Epigenetics

Epigenetics are the biggest potential hurdle to RSCNT. There is a growing body of thought within the research community that cancers can form without mutation of the nucleotide. That is, that the DNA can be mutation free, but that there can be problems with the expression of the genes, resulting in cancer formation. Histones are a commonly cited possible cause, although with my idea – replacing whole chromosomes – you would also be replacing the histones that the DNA wraps around. mRNA and other such epigenetic features of the cell are much more tricky to work around.

I would really like to see my proposed experiment done with one of the HeLa cell lines. Granted, I doubt that there is any surviving non-cancerous DNA from Henrietta Lacks in existence, but any donor DNA ought to work. SCNT has proven that there is not a rejection risk with nuclear material like there is with organs in the body. Current theory suggests that epigenetic alterations last through cellular division and will be present for multiple generations. If that is true, then we would expect that – at least some of the time – RSCNT would not immediately cure cancer.

Although, I would like to say that they have yet to convince me on the whole epigenetic-caused cancer theory. It seems a lot like a chicken-and-egg kind of thing. I think a lot of the so-called epigenetic cancers with no apparent nucleotide mutation might be the result of a poor understanding of the workings of the entire genome. Defective genes that they don't even realize are defective because of their prevalence among humans could be the culprits, and it might take the interaction of several of them to cause epigenetic alteration.

That said, if current epigenetic theory holds true and epigenetic alterations last for multiple generations (but are eventually corrected once the DNA is non-mutated) then we would expect that cells treated with RSCNT would continue to behave like cancer cells some of the time and, as a group, individuals would begin to behave like healthy cells after undergoing mitosis until, gradually, all the cells eventually rebuild their epigenetics based off of their DNA. If epigenetic alterations prove to be a permanently inherited trait, or last for too many generations for RSCNT to be able to operate as a sole cure, then other methods will have to be used in conjunction with it.

One method, off the top of my head, is my organelle idea from earlier. Since cancer cells share the characteristic that their DNA continues to mutate (and the mutations get progressively worse as the cancer cells divide), then any method of comparative analysis would weigh in favor of healthy cells on the borders of the cancer. If you create organelles that do the same thing, but for the epigenetic components of the cell, then they would work their way from the outside in, correcting the cancerous epigenetics until all the cells were healthy. However, in this case it might be easier to design an organelle that causes apoptosis when dramatic changes are detected. That's just a thought, though. One of the limitations of this idea is that any check on epigenetic alterations would have to take into account the type of cell that it is in, since epigenetic differences between cells determine cellular determination. We would not want to create something that turns liver cells into epithelial cells simply because they are neighboring.

The nanorobotic idea that I mentioned earlier may be an even better solution, but the needed advances here would make it even more out of reach for now. The nanobots would actually have to possess some form of intelligence so that they can identify what kind of cell that they are in and only compare cells of the same type. Then, you would have to give them the ability to determine whether or not a cell is cancerous and how to react to that. Horribly complicated, and hopefully unnecessary if epigenetic alterations prove to last a few generations or less.

37 trillion

There are a crap-ton of cells in the human body. Thirty-seven trillion or so (the number given widely varies depending upon the source – a lot of sources you find say between 1 and 10 trillion while others say 100 trillion – I like the 37 trillion figure because it seems so precise). That's 1.7 quadrillion chromosomes if each cell only had one set – which they don't (stupid muscle cells and your multiple nuclei). Assuming 6 grams of DNA per trillion cells, that works out to 222 grams for a human, or almost half a pound of DNA.

That's not that much material, and the actual cost of the component elements is trifling (and really, if you can figure out a way for your DNA-removal vessels to collect the DNA that they have removed into one location, you can break it down and use it as the building blocks of your next customer's genetic overhaul). The problem, though, is how do you replicate a healthy genome that many times over without error? Industrial means will have to be created to accomplish the task. I would assume that some kind of engineered organism would be used at first to rapidly multiply selected DNA. Perhaps using some sort of SCNT therapeutic cloning would be the best near-term solution to the problem.

In therapeutic cloning, a donor nucleus replaces the nucleus of an egg cell, creating an embryo. Once it has multiplied enough that the cells begin to differentiate, its cells are split apart and, in theory, each cell goes on to grow into the tissue that it had differentiated into. So, to use it as a means of farming healthy DNA, you would conduct SCNT many times over, perhaps forming hundreds of embryos, then split the cells apart and you would, in essence, have healthy DNA growing on an industrial scale. Once the mass of the tissues has exceeded the mass of the person with cancer, and quality control genome sequencing of the tissues has ensured that the replication was without error, the DNA would then be extracted by the vessels (either organic or mechanical) used for the RSCNT process.

That, of course, is an ethical nightmare waiting to happen. The goal should be, then, to be able to mechanically create DNA. It should be hoped that in the future machines can be made that can assemble, without error, a sequenced genome. So, after analyzing the healthy DNA, it treats the information as a program that it then follows to create artificial DNA. Artificial is really the wrong word. DNA is simply an incredibly complex molecule. I mean artificial in the sense that it would be compiled by machines instead of copied within the cell as part of the long chain of life and evolution. The machines, of course, would also have to create the histones that the DNA winds itself around.

Time

Any method that is developed will have to be industrialized to the extent that it can quickly go from extraction of the template DNA to fully replicated and ready for RSCNT for that individual before they reach a point that their tumors are inoperable and debilitating them to the point at which survival is unlikely. Some people with certain cancers can last years before they find themselves on death's door. Other people with more dangerous cancers (I'm thinking of pancreatic cancer in particular) may have a handful of months.

Ideally, we would eventually reach the point at which a person learns that they have cancer, their doctor takes several healthy tissue samples, their DNA is extracted, replicated, collected by whatever means of RSCNT will be used, and then are treated within a few days, long before their cancer has any effect on their quality of life. That is the ideal, but it is a far-reaching ideal. We're talking about a massive investment of resources to be able to provide that quality of care to every single cancer patient. Over 12 million people are diagnosed with cancer worldwide every year. Purely mechanical means of DNA creation may be the answer.

The germline

If you develop a means of replacing the genome inside of every single cell in the human body, you quickly run into a problem. That problem, of course, is the germline. Replacing the 46 chromosomes inside of each nucleus of every somatic cell in the body simply gives those cells a fresh set of chromosomes. Replacing the 23 chromosomes inside of each germ cell (sperm or egg) in the human body with 46 chromosomes would be disastrous. I expect that if it was to happen to the egg cells in a woman's body, that she would find that all of the eggs inside of her would be fertilized at once, causing a different sort of cancer. For men, I expect that they would simply have sperm with 46 chromosomes until spermatogenesis completely replaces all of the sperm within his body. It shouldn't do anything to the man, but means would have to be put into place to ensure that he did not fertilize any women with the defective sperm. It's hard enough to convince people with HIV to wear condoms. Throwing numbers at men may be a tad too much.

Then, there is the fact that you, at least, want the RSCNT process to affect the spermatogonium – the cells that create spermatocytes. The reason is that, while rare, germ cell cancers are possible, and, in any event, if you're ensuring that every cell in the body is free of mutation, you especially want to ensure that the cells that are passing on traits to the next generation are as free of mutation as possible. So, the problem is how to create a RSCNT process that only affects somatic cells.

The answer may come from the field of human gene therapy. Gene therapy has a lot in common with my general idea, but they are only working to alter a single gene or a few genes inside of each cell instead of replacing whole chromosomes. The important point, though, is that several jurisdictions across the globe are moving to ban gene therapies that are inheritable. Now, since fear mongering does not stop research, but simply sends it along new tangents, they are undoubtedly working on means of conducting gene therapy that only target somatic cells, leaving germ cells alone. Whatever their solution will be will certainly be applicable to RSCNT.

Approval

To date, there has not been a single genetically modified animal approved for human consumption by the FDA. That is after decades of research and unfathomable amounts of money spent. That isn't to say that several safe varieties have not been created. They have. But public fear and backlash combined with bureaucracy have prevented their approval.

Given our current state of technological development in the areas of genetic modification, I would estimate that if several biotech firms were to immediately start development on RSCNT (and that it's even possible), that it could be developed and ready for human testing within ten to fifteen years. I feel that's a very conservative estimate based upon trends in the field. After that, I foresee at least a decade-long battle, probably twice as long before human testing is approved in the United States. And then we have the extremely long approval process...

If RSCNT is ever to be used in the United States, then the hurdle of our precautionary principle-based bureaucratic overlords will have to be overcome. From the moment that a cure is developed until it is actually available for sale millions will die that did not have to. There's no getting around that. Hell, if RSCNT was developed *today* I would probably be dead by the time that the FDA let it out onto the market – from one of my many habits or from some of the many things that I have exposed myself to (I grew up on a farm and ranch. Exposure to chemicals such as pesticides and fertilizers was an everyday part of life. That continued during my second job, when I worked in a factory as a sanitor that handled caustic cleaning agents, not to mention that asbestos warning signs were a common sight throughout the plant. And then there are the stupid things that I did, such as breaking open and collecting mercury from old appliances when I was a kid – which might be part of the reason that I don't worry excessively about my smoking habit – I don't expect to exceed the median life expectancy anyway).

My recommendation to the biotech firms, however, would be to apply for approval through the FDA while simultaneously finding another country (probably in the third world) that would allow them to immediately begin human testing. Perform the cure for free for a few million people. Build a history of effectiveness and safety that is so irrefutable that the American public's demand for its introduction will overcome the hesitation of the FDA and the left and right wing nuts (and, in fairness, the average American that believes anything they read on social media) that vociferously oppose GMOs.

Possibilities

Introduction

If we take it as a given that reverse somatic cell nuclear transfer is possible, then it opens up a plethora of possibilities. Sure, curing cancer is the catch – the draw that will entice investment in the idea, but there is so much more that can be done with it! The following chapter is for the hopes and dreams of a future in which humanity is able to conquer many of the seeming-mountains that have stood in its path and rise above them.

Karyotypes

Correcting chromosome abnormalities were one of the very first applications other than cancer curing that occurred to me after the idea of RSCNT popped into my head. Both numerical and structural chromosome abnormalities can be addressed using RSCNT, with the addition of donor chromosomes, depending upon the abnormality. Most abnormalities happen as an accident in the creation of the sperm or egg cells and are thus present in every cell in the body, although some happen after conception has occurred, causing mosaicism, in which different cells in the body have differing numbers or structures of chromosomes. I will briefly discuss how I believe RSCNT can be used to treat some of the most common chromosome abnormalities.

Klinefelter Syndrome

The most common chromosome abnormality is Klinefelter Syndrome, in which a man has at least two X chromosomes and a Y chromosome, for a typical karyotype of 47,XXY (or 48,XXXY – etc.) instead of the normal 46,XY. This disease afflicts 1 to 2 males per 1000, and its effects are usually sterility, enlarged breasts, taller than average height (which is hilarious when you consider that height is the attribute that study after study has found women to value most in a partner – her tall man may very well be not much of a man), reduced muscle mass, less facial and body hair, broader hips, weaker bones, and they are more likely to suffer from diseases that afflict women. The solution, as I see it, is to take the DNA of an afflicted male, remove his excess X chromosomes until the sample has only one X and one Y, then follow the RSCNT process, replacing the DNA in his body until every cell's nuclei are 46,XY. You would want to correct the condition as quickly as possible, but for this disease correcting it before the onset of puberty is probably good enough. Puberty, of course, is when most of the symptoms of the disease become apparent, although there is an incidence of speech and learning disabilities that occur during childhood, so the earlier you catch it the better – you would want to give their brains the maximum amount of time as possible to develop with normal 46,XY controlling DNA. I would assume that correcting it before puberty would correct the most common symptoms, such as infertility and the development of feminine traits.

Trisomy

There are various types of trisomy (having three of any one chromosome instead of the normal two), with the most common being Down Syndrome (trisomy of chromosome 21), Edward's Syndrome (trisomy of chromosome 18), and Patau Syndrome, (trisomy of chromosome 13). These diseases are extremely debilitating, Edward's Syndrome being the most, with horribly few of the live births suffering from it living past one year.

A skeptical reader might have noticed that I called this book *One Cure* but outlined a 'single cure' that is more accurately described as individualized medicine since the idea is that each treatment uses DNA specific to that person. Well, hypothetical reader, I simply did not elucidate earlier on how you can have a single cure because, in most cases, I believe that it would be unethical. With trisomy, and other horrendous chromosome abnormalities that are present at conception and begin to affect the fetus' development immediately, I believe that using a true 'single cure' is ethical.

My idea is as follows: perform RSCNT, but use a standardized set of DNA that is not from that person, with a normal set and number of chromosomes. That way you would have a treatment already ready to go even before the diagnosis is made. Kind of a 'standard' cure that can be ready-made. Here I am advocating that karyotype screening be conducted on all fetuses as quickly as possible after conception. As soon as an abnormality is identified, a sample of DNA can be taken from the fetus and the ready-made RSCNT procedure can be performed. Of course, a method of conducting RSCNT in utero will have to be developed that selectively replaces the DNA of only the fetus and not of the mother.

The earlier you identify the trisomy in the fetus and perform RSCNT, the better. With the normal set of DNA replacing their original set, the children should begin to develop normally, so if you catch it within a few weeks of conception, they will develop normally from that point onward. If you can test for and catch it before the heart and brain are fully developed, even better – the resulting children would probably be almost impossible to discern from children that had never had trisomy.

Of course, you would not want to keep the standardized DNA in the child. Here's where the ethical implications that I discussed earlier come into play. You want the child to develop along the genetic path that they would have gone down had they never been diagnosed with trisomy. Also, you don't want the human population to begin to bottleneck through a single combination of

DNA. Think about that: in just the United States, around 4 million children are born every year. Down Syndrome alone occurs in around 1 in 1000 births. If every child testing positive for Down Syndrome was treated with a standardized DNA RSCNT and then we left it at that, then you would have around 4,000 genetic clones produced as a result every year. And that's just for one disease. Hell, adopted siblings somehow manage to find each other and fuck all the time. Throw several thousand genetic clones in the mix every year and eventually the inbreeding rate would become ridiculous.

The solution, then, is to sample the fetus' DNA prior to conducting the standardized RSCNT treatment. Take the sample, and remove the excess chromosome in the fetus' DNA. Then, multiply the DNA through whatever the normal means are, and reintroduce it using *in utero* RSCNT. The earlier the better. You want the child to actually begin to develop according to the traits of its parents as early as possible. I can only imagine the social implications that it would bring if the standardized DNA used encoded for a child of African descent was used to treat a Caucasian couple's fetus, and the child was born before its corrected DNA could be reintroduced and control its appearance.

Monosomy

Full monosomy is only seen in Turner's Syndrome (45,X0 instead of 46,XX) in humans, which is when a child is born with only one sex chromosome, which is the X chromosome. All other forms of full monosomy result in death during development. Partial monosomy is sometimes survivable, such as Cri du chat syndrome (partial deletion of the small arm of chromosome 5) and 1p36 deletion syndrome (partial deletion of chromosome 1). I expect that treatment of Turner's Syndrome would be accomplished by testing the existing X chromosome to determine which parent it came from, then either adding another copy from that person to double the number, or – preferably – taking an X chromosome from the other parent and adding it to the person's DNA sample. Then, you can perform RSCNT (as early as possible being the key to all chromosome abnormalities – for the sake of no longer being redundant we will just assume that from here on out anything that can be performed in utero should be – and as early as possible) to bring the fetus' karyotype to 46,XX. The partial monosomy disorders can be treated in much the same way as Klinefelter Syndrome, the difference being that you would remove the chromosome that was partially deleted and replace it with a good copy from either the person's parent that it should have come from or another donor (or even double the patient's good chromosome, but that would cause its own host of problems, particularly if that chromosome has any recessive genes that code for genetic disorders – in lieu of their parent's chromosome, a stranger's donated chromosome may be the best option).

Mosaicism and Chimerism

Mosaicism occurs when the chromosome abnormality occurs after conception. Not every cell in the body will have the same set of DNA. The trisomy diseases are mosaic rather often, as well as monosomy diseases, such as Turner's Syndrome. 30% of people suffering from Turner's Syndrome are mosaic. Chimerism is similar, in that the person afflicted has cells with different DNA throughout their body, but its origin is usually that two fertilized eggs merged together in the womb, developing into one person from two sets of DNA. Of interest here is that it is the cause of a lot of intersex disorders.

Mosaicism and chimerism can both be corrected by deciding which set of DNA that the afflicted fetus has that you want to keep and using that set for RSCNT. In the case of trisomy and monosomy mosaicism, it would be an easy choice: whatever set has the normal number of chromosomes. In the case of chimerism you're faced with a more difficult choice: which healthy DNA set do you want to use? In the case of 46,XX/46,XY fetuses in the womb, it would be a choice of what sex the parents desire the child to be, although it might end up that children that started as intersex chimeras have a greater chance of 'normalcy' if one is chosen over the other due to the hormonal effects of the presence or absence of an Y chromosome early on. Non-intersex chimerism decisions could be based upon which set of DNA has the least number of genes that code for genetic disorders.

Structural Abnormalities

So far, we've only dealt with disorders that have too few or too many chromosomes, or in which the person's DNA is not the same throughout their body. There are a number of disorders in which a problem occurred, usually during the creation of the sperm or egg, that changed the structure of one of the chromosomes. A portion of the chromosome could be deleted, such as in Wolf–Hirschhorn Syndrome and Jacobsen syndrome. Part of a chromosome could be duplicated, such as in Charcot–Marie–Tooth disease. A portion of one chromosome could transfer to another chromosome, such as in a Robertsonian translocation (which scared the shit out of me as I read about it – it can be asymptomatic and can be inherited). There are other abnormalities that can occur, but it isn't important that we go through them all.

What is important is the way in which they can be treated. Seriously, read about the disorders that I named. Fuck that shit. With an abnormality that changes the structure of a single chromosome, it would be a simple matter of replacing that chromosome with a normal one from their misrepresented parent (or – in the case of inherited abnormalities – with a normal one from a family member on that side, preferably, or a donor). An abnormality that changes the structure of several chromosomes, such as with a translocation, would have to involve the replacing of all the affected chromosomes. Same general idea, however. Conduct fetal RSCNT, and the problem is mitigated.

One Final Thought

An interesting statistic that I came across when researching chromosome abnormalities is that in Europe 92% of pregnancies that test positive for Down Syndrome are aborted, while in America 67% are. And that's just the statistic for one disorder – one that, while a tragedy, you still wind up with a child that is still able to communicate (albeit their average adult IQ is only 50 – equivalent to that of a child) and live a life that is describable as human – one of the milder chromosome abnoralities. A good deal of what I have been doing here is making the case for developing *options.* We could discuss the political, social, religious, and economic reasons for that disparity all day long. We are not going to here, though. It is undeniable, however, that parents that raise a child with a chromosome disorder, such as Down Syndrome, are currently consigned to a parenthood that is much more difficult and expensive for all involved, including society as a whole, which has to use extra resources to accommodate the afflicted person's condition (although – going on a tangent here – that doesn't necessarily mean that society as a whole *should* use its resources or be *compelled* to use its resources to care for anyone, no matter how disabled. Simply because I argue that the genetically disabled are more resource intensive for everyone involved does not mean that I condone a social concept of community care or even think that it is a necessity).

If developing an option that would allow (for those treated early in the womb) the afflicted to lead more normal or indistinguishable from normal lives, even if the abortion rate remains unchanged, then it would reduce the economic burden of those parents that choose to keep their children – and treat them with RSCNT – and for society as a whole. Of course, this would also make the case for near-ubiquitous early testing for chromosome abnormalities and karyotyping. And would probably change some of the nature of the abortion debate (as applies to allowable exceptions to restrictions). But we're staying out of politics and morals in this section.

Genetic Engineering

Gene Therapy

It occurs to me that RSCNT might be the perfect vehicle for gene therapy. As of right now, researchers are having a problem with ensuring that targeted genetic changes using the various experimental gene therapy techniques such as CRISPR, TALENs, and guide RNA are ubiquitous across the entire test subject. Current potential applications hope that *enough* cells can be modified with a desired gene to reduce the effects of a genetic disorder or counteract it.

I see it as a problem of complexity. The various vectors used to enact gene therapy as it currently stands have to enter the body, get inside the cells, target the desired genes on the correct chromosomes, make their changes, and repeat throughout the entire body. I believe that if you first remove the sample DNA from the body, then apply your desired gene therapy to the DNA while it's outside of the body (*in vitro*) and to only the single set of DNA, then you have a much greater chance of success. You keep trying until you get it right. Once the DNA is modified to your satisfaction, then it can be reinserted into the body using RSCNT. The complex actions – DNA modification – would have been done outside of the body, leaving the simple actions – replacing all the chromosomes – would be done inside of it.

And then, of course, if you can figure out how to produce artificial chromosomes mechanically – as I discussed earlier – then it's a simple matter of changing the programming to reflect your desired results. 'Oh, this genome scan that we've uploaded into the chromosome creator shows that the patient codes for a defective *BRCA1*.' The tech types a few buttons, clicks the mouse a few times... 'And now they do not.'

Genetic Disorders

The most obvious and immediate use for RSCNT as a means of implementing gene therapy in humans is the eradication of genetic disorders. The over 4,000 single-gene diseases in humans could be cured in two generations using early, fetal genetic testing and RSCNT. Why two generations? Because, as I noted earlier, RSCNT will undoubtedly have to be devised in such a way to leave women's egg cells alone. Therefore, every adult woman with a genetic disorder that is treated using RSCNT will still have a chance of passing that disorder on to her children. By then testing her children and performing RSCNT on them quickly enough – in this case before her female children's oogonium turn into oocytes during the third trimester – you will preserve their fertility while removing their ability to pass on genetic disorders. That is, by treating female children with RSCNT before they develop oocytes, their egg cells will develop with the desired DNA.

Realistically, the polygenic and multifactorial genetic disorders could also be removed from the human genome in two generations as well, but the single-gene disorders are the low hanging fruit that are out there. Note that I am talking about autosomal dominant and recessive genetic disorders as well. What diseases are we talking about? Well, the most obvious to me is sickle-cell anemia. After that, Huntington's disease, cystic fibrosis, albinism, and even the sex chromosome-linked diseases such as hemophilia, color blindness, and male pattern baldness. My point being, that there are thousands of diseases that can be eliminated with relatively minimal effort.

The more complex genetic disorders will require more research into their causes, but I do not doubt that their genetic causes can be identified and counteracted using RSCNT in conjunctions with gene therapy. We're talking about the diseases that 'run in families' but don't follow the Mendelian pattern that they teach you in school (the bit about the monk, peas, height, and a grid square). We're talking about obesity, heart disease, asthma, poor eyesight, and being a ginger. The polygenic diseases tend to need both their genetic component, with several genes needing to be present for the disorder to take place, along with an environmental factor such as the presence of the Devil being the environmental factor for gingers (you'll have to pardon me, readers that don't understand the sense of humor that I've sprinkled throughout this book – my daughter is a redhead – I imagine someday that she'll read this and the joke will be worth it).

At this point, we're talking about removing the source of huge economic burdens from society. Hell, the money that is wasted fighting cancer itself is unfathomable, but now we're removing the burdens of obesity and heart disease. If I haven't already made the case for pouring ungodly sums of money into RSCNT research yet, then that ought to do it.

Preferred Human Traits

Quick, what advantage do indigenous Mexican natives and the Inuit have over the global population at large? The answer is that it is much more likely for them not to have wisdom teeth. 'So what?' you may say, and I admit that wisdom teeth are something that no one thinks too much about unless they are told that they have to have them removed, but the numbers might surprise you. 72% of the population develops impacted wisdom teeth, quite often having to have them be removed. Less than 2% of people with wisdom teeth are able to maintain them without developing caries or periodontal disease by the age of 65. They are clearly an undesirable trait to possess. However, a mutation occurred in a population of humans thousands of years ago (before humans migrated across the Bering Strait, which is why a lack of wisdom teeth is also common among Asian populations) and spread – mainly in the Americas. This mutation exists within the human genome naturally and is what I call a preferred human trait. You are better off if your DNA codes for a lack of wisdom teeth than if you have them.

A lot of the controversy that surrounds human genetic engineering is over the idea of designer people. Granted a lot of this argument has an economic basis: people are afraid that the rich will be early adopters (like they are with every technology, such as the automobile, television, cell phones, and electric cars) and foretell doom if genetic engineering is allowed to occur with money changing hands. The idea is that the rich will be advantaged and that it will be horrible. My position is: so what? After they've milked the rich for all that they will pay, in a market-based economy the price will drop so that the developers of treatments can earn money from even the most average of consumers – just like they did with every other emergent technology.

I started by talking about wisdom teeth because it's an example of the good that designing people can bring about. There are plenty of traits within the human genome that some people possess that have a clear advantage over traits that other people possess. Morton's Toe, for instance. If it wasn't as common as it is

– afflicting roughly a tenth of the population – it would be considered a disease instead of a normal variation of the human genome. Poor vision is another good example.

Certainly, if RSCNT is used as part of genetic engineering, we will begin to see all of the things that fear-mongers warn about: I expect that we'll see more tall people, more blue and green eyes, more blondes (although, I have never understood the fascination with blondes – raven-haired women with light eyes are clearly superior), and the elimination of redheads, but we can also remove the traits that hinder people that possess them, such as fucking wisdom teeth. Think about that: a future world in which having to undergo surgery because the evolution of your brain has outpaced the size of your jaw will be a thing of the past.

Intelligence

I went to what is considered a pretty good public school in Texas. It was rated 'Exemplary' while I was there (Texas has/had something like three different good ratings that could be bestowed upon schools – it wasn't an inner city school is what I'm saying). It may still be. Dunno. Don't care. Anyway, what struck me was just how… *stupid* a large number of my classmates remained after receiving the same education that I received. We're talking about a rural school that served grades sized from about 40 to 70 max, depending upon the year. The largest classes had about 25 students per teacher. There was no great socioeconomic division. We were all country kids. There was no great racial divide. There were a handful of minorities in each grade, but they seemed to vary as much as the other students, with a couple of them being exceptional students, most average, and a couple being stupid. There also wasn't some large transfer population problem skewing statistics. I would estimate that around three quarters of my grade went to our school from kindergarten all the way until they graduated high school.

No, what struck me was that after receiving the exact same education that I received so many of them would struggle with what I would term basic skills required of a good citizen. Senior English was the point at which it really struck me just how unprepared some of my fellow students were for the life ahead of them. So, as I'm certain every student does (except for those weird liberal schools where they don't do anything that makes people feel uncomfortable or highlight exceptionalism), we would take turns reading from a book out loud during class. Some Shakespeare, *The Great Gatsby*, and I think *Brave New World* and *Animal Farm* were two of the books that year. Irrelevant. The point is that someone for whom English is their native language, and that has received thirteen years of instruction in how to read it, *should* – at that point – have achieved something of a mastery of their own language. And they had not. We would bear through student after student that struggled with sounding out the simplest of words. It was as if we had been transported back in time to the beginning of our education and were struggling with *See Spot Run*. These same people would then struggle their way through the most basic of math and science classes (and I'm not convinced that they ever really understood what was being taught to them), and at the end received a diploma.

At least up until they read this (Hah! Read.) almost all of them are friends with me on Facebook, so I've been able to see where their lives have taken them in the dozen years since we were let loose upon the world. And they're breeding. And voting. Functionally illiterate people. I'm not entirely certain what part of that bothers me the most.

That said, I believe that RSCNT can be used as a means of genetic engineering to uplift the entire human race. That term is, of course, stolen from David Brin's *Uplift* series of books. He also wrote *The Postman,* among other things. Great books. His blog, though, is liberal propaganda. Tangent ended.

I like to think of intelligence has having many components that can be related to computing. You have a storage and recall component, analogous to a hard drive. The best variety can store the most information and more readily recall it. You have an ability to juggle multiple thoughts at once, analogous to RAM. The greater your ability there, the more things you can think about at once (beyond breathing and walking without choking on the gum in your mouth). Then, you have a processor. Some can only ever regurgitate information that they've experienced and will stall as the operations get too difficult, while others are capable of novel thought – of being true thinkers. Finally, you have an analogue of an operating system (maybe hardware based OS is a better analogy) that guides your thinking, and I believe that here we have the most important component to intelligence, because *curiosity* appears to be the component that the greatest number of people lack.

Intelligence, raw intelligence, definitely has a genetic component. Almost certainly several components. We need to identify what those genes are and give them to the world. Certainly, giving people a greater ability to store, hold in their mind, and process greater and more complex amounts of information is a good start, but we need to give the world curiosity. It is the spark that *drives* a patent clerk to work on his theories in his free time, a printer to fly a kite in a storm, a farmer to think about what makes men equal, the difference between a child reading everything that he can get his hands upon and another child who is given every opportunity to excel and fails. It is the power to change the world.

Transhumanism

All of the genetic meddling that I have so far discussed has dealt with giving people genes that they do not possess but other people do – that is, only using DNA that is human. I'll go off on a tangent here and talk about what is known as the Heinlein Solution, named after Robert A. Heinlein, who in his book *Beyond this Horizon* laid out the idea that in the future couples could select the sperm and egg of their choice to make a child. So we're talking about genetic engineering, since you would be selecting a child to be as free of disease as possible and with the most desirable traits, but it would have been a child that the couple could have potentially of had anyway – although it would have been so mathematically unlikely that it becomes near impossible. My above discussion is much the same. We've been talking about modifying humanity in such a way that it would have been possible to do with selective breeding. Call it the All-Human Solution to genetic engineering.

But, then, think of how much there would be to gain if we stepped outside of those restrictions! My personal favorite idea (which is expressed on David Brin's blog) would be to add the chimpanzee genes for different ligament attachment points (along with other muscular advantages) to our genome while keeping our genes for fine motor control (we have more neurons that control our muscles). The innate strength of a chimp, but with the control of a man. It's a wonderful idea. The animal kingdom is filled with possible adaptations that might benefit us. Tetrachromacy is one (and yes – I realize that a very small number of humans have been shown to be tetrachromats, but what I'm talking about is the ability to see the visible spectrum *plus* ultraviolet light). The exact benefits of that one elude me for right now, but imagine how it would feel to have a whole other spectrum of light opened up to you – to see a flower the way that some insects and birds do (or as Geordi LaForge would). Here's a good one: limb regeneration. Or at least the ability to regrow lost digits. How about the ability to manufacture vitamins within our bodies that we currently cannot? Most other mammals have a gene that produces L-gulonolactone oxidase, an enzyme that can make vitamin C, but a mutation deactivated our gene for it. That's a simple fix right there.

Then, there are modifications that we could make to ourselves that would improve our health by changing how the body develops. Take, for instance, the appendix. It's vestigial, does nothing for you (fuck off science and your constant search for the benefits of it – people live perfectly normal lives after they're removed – you're just enabling vestigialism), then one day you wake up with severe abdominal pain and can die if it isn't removed. I say we remove the genes that code for it out of our DNA.

How about the post-mitotic nature of our muscle and nerve cells? I fully believe that we should tinker with the genes that control that and figure out how to make them divide and replace neighboring cells that have been destroyed. The fact that they do not undergo mitosis is a major contributor to the onset of overall debilitation that we call old age, which brings me to another point...

Cellular Senescence

The fear of cancer is one of the major reasons that we have not experimented with lengthening of telomeres in humans. I've already discussed the Hayflick limit and how it normally leads to senescence once a cell reaches it. The fear that I speak of is of course that if you remove the Hayflick limit, suddenly *all* tumors become cancer – they just keep growing. However, in the world that I have described – one in which cancer is something that is curable with an injection a few weeks after it has been diagnosed – that fear need not apply. I see two options for dealing with the Hayflick limit, which I will call the innate solution and the *in vitro* solution.

The innate solution is to genetically modify humans so that our cells produce telomerase, probably during mitosis. This solution isn't so much about adding any foreign genes so much as it is pinning down how to make *all* cells do it. Male germ cells, stem cells, and embryotic stem cells all do it, but the somatic cells lose the ability to. This also ties back in to my idea of engineering post-mitotic cells to be able to divide. Aging, and its associated degeneration of the body, can be seen as a breakdown of homeostasis in the body. If we engineer the body to maintain homeostasis – by continuously replacing cells that no longer function as they should – then the degeneration of old age can be stopped.

The *in vitro* solution is that the lengthening of telomeres can occur outside of the human body during the RSCNT process. For example, if a biological method of DNA replication is used, then the DNA would be exposed to telomerase throughout the replication, resulting in telomeres that are lengthened. Probably substantially lengthened. However, if a non-biological method of DNA replication is used – such as the creation of artificial DNA through industrial means – then the telomeres can be lengthened to spec, I imagine, for a surcharge – "For only $1,000 extra dollars we can add ten years to your expected lifespan. Would you like to purchase our telomere extension at this time?" The end result is that the telomeres of the DNA being placed into the cells during RSCNT would be longer than the ones that they are replacing – the patient would have longer to live without dealing with the effects of cellular senescence.

Anti-aging therapy

The net result, then, is that RSCNT can be used as an anti-aging therapy. In particular, the *in vitro* solution can be used as a *continuing* therapy. The telomeres can be lengthened enough to create satisfactory results, but not so much that he patient can then go decades without needing another treatment. They could set it up so that the patient has to return every five or so years for another RSCNT therapy to continue to reap the anti-aging benefits.

Of course, now we're getting into the ethical concerns that sceptics of progress use to stop research into promising fields. After all, while the rich would surely be able to afford RSCNT as an ongoing therapy, the poor surely would not. See above my argument concerning emergent technology for my opinion on the matter. That said, the innate solution would be a more ethical answer to the problem, since the modification of the genome would be heritable – with socially selective breeding (i.e. people make a conscious attempt to only have children with people that have the modified genes) eventually all of humanity would have telomeres that do not have a Hayflick limit.

One final argument against that, though, would be dark ages. If you modify the human genome so that telomeres do not shorten, then you risk the end of the human species in a dark age. Look on your body. Do you have any moles? Congratulations, you have a tumor that eventually stopped growing. Without the Hayflick limit, every tumor that you have ever had would have become cancerous. If we were to enter a dark age in which RSCNT was no longer available, then the likelihood of people dying off from cancer before being able to reproduce would be dramatically increased. So then, if the innate solution is used, then it must also be coupled with increased DNA repair abilities such as the ideas that I mentioned earlier involving heritable organelles that correct for mutations.

Transplants

I thought about not including this possibility. After all, presumably in a world in which RSCNT has been perfected therapeutic cloning would also have been perfected, eliminating organ transplant rejections. However, it occurs to me that we would probably still have a need for emergency transplants from donors as a means of lifesaving – either to replace organs in individuals that have had catastrophic damage occur to their bodies and have too little time remaining for an organ to be grown, or to replace defective organs in individuals with genetic disorders that have yet to be treated with genetic engineering.

Acute rejection of transplanted organs occurs to some degree with every transplant, which is why immunosuppression is so important – with the exception of transplants between identical twins, in which acute rejection does not occur. That's our first clue. Identical twins have the same DNA. They produce the same antibodies and their DNA encodes for the same major histocompatibility complex (MHC) surface molecules on the outsides of their cells.

So then, if you can transplant an organ without hyperacute rejection occurring – which happens within minutes and is immediately life threatening – then you could treat the patient with RSCNT to eliminate organ rejection. That is, during RSCNT the patient's DNA would be replaced, along with the foreign DNA of the transplanted organ – to include its antibodies. The cells would then express MHC surface molecules that are compatible with the rest of the patient's body, eliminating rejection and eliminating the need for the patient to undergo continued immunosuppression therapy.

The process, as described above, is a patient-side therapy. There is another possibility, however. Donor-side therapy. In order to reduce the chances of hyperacute rejection from occurring, RSCNT can be performed on the donor. My thoughts on this one, of course, would necessarily be limited to a specific number of situations in which it would be ethical. For example, let's say an organ donor who is otherwise completely healthy winds up in an accident and dies. Perhaps they suffered massive brain damage and their body is hooked up to machines so that their organs continue to function but there is no hope of them recovering. However, let's say that due to them having a unique MHC class or one of the weirder blood types that their suitability as an organ donor is called into question: it is expected that hyperacute rejection will occur with anyone that their donated organs are given to. In that case, performing RSCNT *on the donor* using a standardized DNA with a common MHC class and giving them a useful blood type and Rh factor (such as O negative) would maximize their utility as an organ donor.

For another example of when it would be ethically sound to change the donor's DNA through RSCNT, we would look at consent. Let's say that a child has an immune problem exacerbated by a failing organ – such as a kidney – and they have a willing donor. However, it is determined that they are incompatible with one another. Let's say that risking patient-side RSCNT after transplant is not an option due to the patient's immune disorder (RSCNT using a virus as a vector). In that case, if the donor is willing, then they can undergo RSCNT to match the patient's DNA. Their organ can then be transferred without fear of rejection. Once the donor has healed, then they can once again undergo RSCNT and return to their former DNA.

Fertility

A procedure that could be derived from RSCNT that occurs to me is as a means of conducting *in vitro* fertilization. During intracytoplasmic sperm injection, the egg cell suffers a lot of damage from the needle used to insert the sperm cell. Quite often multiple egg cells are injected with sperm until one begins to successfully divide and form a zygote. This can have its own moral and ethical implications, depending upon how many failed zygotes have to be discarded, or what are to be done with the extra ones when too many are successful. The same applies to non-invasive (to the egg cell) procedures such as when the egg cells and sperm are simply allowed to mingle in a petri dish.

There are a number of reasons that intracytoplasmic sperm injection is chosen as a means of accomplishing conception. They're unimportant. What is important is that when it is used, it is because it is necessary for medical science to help the natural process along. The problem is that it is so traumatic to the egg cells. Therefore, you could modify the viruses or nanobots used for RSCNT to carry their cargo (the 23 chromosomes taken from a sperm cell) and deposit it into the egg cell without destroying the chromosomes that are already there, bringing the total to 46 and causing conception to occur. Granted, the vessel used would also have to mimic the complex interactions that occur when the head of a sperm cell enters the cellular membrane of the egg cell, but that – like a lot of things that I have glossed over – are details that someone else can figure out. I've been providing concepts, not necessarily solutions.

Since I've already shown a compulsion to name my ideas, we'll call this one germ cell nuclear transfer, or GCNT. The key (and why this idea is a better solution than using a needle) is that the vector used for GCNT would be many times smaller than even the smallest needle that is used now, resulting in far less trauma – if any – to the egg cell. Since there is little or no trauma, you should – in theory – only have to mess with one egg cell and the material from one sperm cell, reducing the aforementioned implications.

Taken further, the same idea can apply to a lot of similar fertility treatments. For instance, it would be an easier way of combining the chromosomes of two women to allow them to make a child. Or, for allowing two men to do the same. Granted, for two men you would first have to ensure that at least one of them is providing an X chromosome and you would have to remove the chromosomes inside of the donor egg cell, but it is entirely possible. It would even be a good way to accomplish what the news has been referring to as 'babies with three parents', where the mother has a defect in her mitochondria, so a donor egg is used, its chromosomes are removed, and then the chromosomes of the mother and father are inserted into it, keeping the donor mitochondria, resulting in a child with genetic material from three parents (which, now that I think about it, is exactly the same thing as making a child from two men).

Then, of course, you can take the idea even further and create a child out of any number of parents from two to 46 (or 47 if we're counting mitochondria). The idea here is that you, presumably as evenly as possible, selectively take certain chromosomes from each desired parent, and so long as you have two of each chromosome in each of the 23 pairs represented at the end, you would be able to make a child. I can really only see people in polyamorous relationships being interested in that idea, but I imagine that they would pay a premium for the complexity involved. This, of course harkens back to my earlier ideas of correcting chromosomal disorders by using donor chromosomes. Here, though, the *intent* would be to create a child from more than two people.

And, of course, there's cloning and therapeutic cloning, but I would hope by now that anyone reading this would have been able to derive those uses for RSCNT from the ideas that I have already discussed on their own. I have chosen what I believe to be the best uses of RSCNT to outline, but there are many others that do not interest me enough to fully flesh them out here. The list of possibilities for RSCNT is very nearly endless, but my patience in writing this is not.

A Leap Forward

Introduction

If RSCNT is pursued as a viable technology, then it will necessitate a great leap forward for humanity. So many technologies will have to be developed or vastly improved that it will change most aspects of human life. I feel that it will be akin to the technological leaps made when man began to go into space. Sure, we can imagine some of the changes that will be made, but it is impossible to imagine most of the benefits that we will gain.

Computers

Storage

Computer memory storage and processing power will have to be vastly improved to deal with RSCNT. So much of the processes that I have described will rely upon computing to be done that there is simply no other way. For instance, take the mechanical creation of chromosomes that I mentioned earlier. I casually stated that the DNA will be scanned and inputted into the computer like a program. However, a single human's DNA is massive, even when represented by only the four common letters used to denote base pairs. Depending upon the way in which the data is compressed, it can be a manageable size – less than a movie file – but still rather large for one person.

But we are not talking about one person. We're talking about a procedure that would presumably be used many, many times over each day. Hell, I was advocating earlier for the genetic testing of every single child as quickly as can be accomplished after conception. That's four million children per year in just the United States. And you would want the data to be permanently stored as well, in case any changes that are made end up being harmful, so each time the patient's DNA is sampled and stored would increase the amount of storage that is needed.

DNA averaging

Another point to make: an idea that I did not mention earlier was DNA averaging. Once you have the means to mechanically create DNA according to a program, you would want to sample DNA from all over the body for each patient. Then, you use the computer to compare each sample against each other to determine what their actual, unmodified DNA is. For instance, let's say you take ten samples from one person. On nine out of ten of those samples the computer finds that there is a certain gene located at a specific location. On the tenth one, however, it finds that a different gene is there. Therefore, it can determine that the majority gene is the one that should be there. By doing that throughout the genome, it can correct the final DNA product to be used in RSCNT against nearly every conceivable mutation – except for ones that occurred very early on after conception.

Processing

So then, we've established that massive amounts of storage will have to be produced. Or rather, that need will drive the creation of massive amounts of storage. However, there will also be a need for increased processing power. Think about what I just discussed. The computer will have to quickly analyze multiple samples of an entire person's DNA against one another, recognizing changes. Alright, that one is simple, but then, once we've combined the idea with the idea of genetic engineering, then it will not only have to look for changes, but will also have to recognize genes that code for less than desirable traits.

That is, it will have to be able to recognize combinations of genes that code for disease, suggest alternatives, and perhaps even flag areas that it recognizes that elective options could be implemented. For instance, eye color or coding for 'normal variations' such as Morton's Toe. The genetic disorders it could be programmed to change automatically, but the elective options it would probably just flag the locations of and allow for manual changes if the customer chooses them. For polygenic disorders, it would have to be able to recognize the combination of several different genes and know how they relate to one another, quite often across multiple chromosomes, and sometimes while taking sex into account. We're talking about a sophisticated amount of processing power here, again, for only one patient. When you scale it up to the size of the population that it will serve, and account for how quickly that it has to be done, then you realize that extremely powerful and fast processors will be required.

Software

Then, of course, we have the programming itself. It will have to be able to recognize how the complex relationships of thousands of genes interact with one another as it searches for genetic disorders and abnormalities. We're talking about extremely sophisticated programming. As I stated earlier, there are 4,000 single-gene disorders – and those are the easy ones. It will also have to be on the lookout for cancer genes, for polygenic disorders, for sex-linked disorders, and for new mutations. We're talking about programs that will almost have to be intelligent to deal with what we will ask of them.

Technology

Even if nanorobotics are not used as a vector for RSCNT, we would still need to develop technologies to deal with the incredibly small nature of the molecules that we would be dealing with. DNA is not on the cellular level. It is on the molecular level.

We will have to develop technologies that reliably extract DNA without damaging it, to handle and sort the DNA as it is multiplied, and to repackage the DNA into its vectors. That's a lot of need for micro technologies. If you actually develop nanorobotics as part of the RSCNT process – as I argued earlier is the best way of doing it – then you would have to develop the emergent technology even further.

For one thing, in order to only target somatic cells would require that the nanobots would have to possess the ability to recognize what type of cell it is dealing with, and act accordingly. That implies that they would possess tiny computers with the programming to match. They'll have tiny sensors that scan the world around them, and tiny tools with which to do their job. Then, they'll have to have tiny little motors and tiny little means of harnessing energy. And that's not even mentioning the processes that will have to be developed to actually make them.

While we're on the idea of industry, let's talk about the industrial processes required to conduct RSCNT. Assuming that mechanical processes to artificially create DNA in the form of chromosomes are produced, then we're talking about entirely new technology. Not only that, but we're talking about machines that will have to be developed on a massive scale. As I stated earlier, the number of chromosomes in the human body is simply mind boggling. So the machines that are developed to create DNA will not only have to create something that is on the molecular level, but will also have to accomplish their task many times over quickly and simultaneously so that it can be accomplished in a reasonable timeframe. Then, they'll have to have the capacity to deal with a customer base that numbers in the millions every single year. We're talking a huge investment in capital. It would certainly be a good time to be an engineer.

Research

As I've hinted at earlier, genetic research will become vitally important. Sure, we know of a few hundred cancer genes right now and feel pretty confident that we know how they work, but we also need to know what combinations of genes cause every single genetic disorder that there is. I firmly believe that once you develop the means to reliably accomplish human genetic engineering, then you will drive a market that will demand that it becomes all-encompassing.

Genetic researchers will become vitally important tools that competing companies use to maintain their market share in the RSCNT based genetic engineering market. 'Sure,' one company says, 'the competition can cure cancer with their version of RSCNT, but we can also treat over 200 genetic disorders.' 'Only 200?' says another competitor, 'we can treat over 500 genetic disorders and the ten most common genetic abnormalities, to include all forms of trisomy.' You should get the picture. The companies that invest the most into genetic research will be able to sell their treatment to more people.

In addition – and this is important – the discoveries that they make would be non-patentable. Well, non-patentable so long as the discoveries that we are talking about are identifying which genes are responsible for what diseases. Anything that naturally occurs in nature is non-patentable. That's why the Supreme Court struck down the patent for the *BRCA1* genetic mutations (although interestingly the Australian court system upheld it – so, using that logic, maybe I should 'find' *gold* and patent it, then demand that everyone using my patented, shiny rock in Australia turn it over). So then, we'll wind up in a situation where thousands of researchers are, in essence, working together by competing against each other, dramatically expanding our body of knowledge.

Care

The way that patients are cared for will dramatically change once RSCNT is fully developed. The most obvious difference is that diseases that today can be characterized by hopelessness will be curable. That, in and of itself, is a paradigm shift of revolutionary implications. So much of our health care is today devoted to managed care that it is preposterous. We pour countless man hours into trying different options, such as different drugs, evaluating the responses, and trying different techniques, such as surgery, angiogenesis inhibitors, or chemotherapy, and yet we still wind up at the same inevitability quite often: pain management and end of life care. Think how much different it would be if the initial chest x-ray (or whatever) that identified an unknown mass was immediately followed up with RSCNT and the problem was taken care of. Think of all the time and money that would be saved. Think of the lives that would be saved. Health care really would become health *care*.

Another dramatic change that would occur would be that we would not be so hesitant to use imaging techniques to diagnose problems. Currently the benefits of high-radiation imaging – such as x-rays – are weighed against the risks – which are an increased chance of cancer. Once you have developed a cure for cancer, however, the risks can be mitigated. Patients can receive regular checkups that include imaging of their innards. By developing sophisticated programs that can compare one full body scan to the next, we would be able to identify tumors long before they could be detected now. Hell, with quarterly scans you might be able to identify tumors that are no bigger than a small pea. The doctor would order precautionary RSCNT, and if it was cancer it will no longer be a problem.

In Summary

At the end of the day, I have faith in the idea of RSCNT. Sure, there are a lot of hurdles to be overcome. It will be phenomenally expensive to develop. It will require technologies that we do not have. We'll have to develop new processes based upon science that we currently do not fully understand. We'll even have to change the fundamental way in which we look at cancer and other diseases.

However, I believe that the possibilities of RSCNT can overcome any hurdles that stand in their way. It has limitless applications that will revolutionize the way that we care for the sick. It will create hope where there was none before. It can even win a few battles against that old enemy of man, death itself. So, instead of using the money that is donated to cancer research to affect social change and compel behavior, I believe that we should actually work on a cure.

That said, if it should (and probably will) prove that RSCNT is unviable, impossible, or not a cure, that doesn't mean that we should stop there. Find another commonality and attack it! Even if the idea is invalid, that doesn't mean that its spirit was wrong. Share this essay with your friends. And tell them to share it with theirs. Eventually someone will read it with the means and interest to actually *attempt* what everyone has written off as impossibility – they will strive for one cure.

Afterword

The idea for RSCNT occurred to me on the night of December 29th, 2014. That date stuck with me because the next day I realized that I had neglected to wish the mother of my child a happy birthday (granted, during the past year she had given birth to another man's child – so I feel that I can be forgiven for not caring too much). You see, I am/was in the process of writing a science fiction novel. At the time I was around 60,000 words in and was preplanning a scene that was to take place a chapter or two after where I was at in my writing.

So, I went out on the back porch to smoke and brainstormed dialogue with myself. Normally I would smoke at my computer while writing, but I was at my house that my brother lives in for the holidays and he is kind of a bitch about me lighting up inside. Personally, I feel that if he paid rent he'd have more of a leg to stand on there. Anyway, my goal was to have one character (who learns about the world that they are in as a vessel for the reader) ask another character about whether or not they cured some disease that was prevalent around the turn of the millennium. This would serve as a way of distinguishing to the reader just how much things have changed in my story, set 300 years from now. Now my first idea involved AIDS. I came up with what I believe is a plausible cure for that disease, but it wasn't working in my head.

Character One: "Did they ever find a cure for this AIDS that I'm reading so much about?"

Character Two: "Of course we did, and you wouldn't believe how easy it was! Blah Blah Blah..."

I found that it was too wordy to describe without going into too much exposition.

So, then, I thought to myself, what's another disease without a cure. How about cancer?

Character One: "Did they ever find a cure for cancer?"

Character Two: "Of course they did. They used genetic engineering to replace the cancerous DNA with healthy DNA."

And then I about choked on my cigarette.

There's a curious sensation that comes over you when you engage in original thought. You get lightheaded. Excitement fills you. Your stomach clenches. It's wonderful. I felt so good that I immediately thought of several other cures that I could write about instead (one example: a way to keep blood vessels free of cholesterol buildup). I had realized immediately that my idea was potentially worth incomprehensibly vast sums of money, and I didn't want to give it away.

Over the next several days I researched frantically. At first, my research focused upon finding out if anyone had ever had the same idea as me. Either they had and never told a soul, or they have and for some reason it's buried. Either way, good for me. Then, I focused upon monetizing the idea. I looked into patents. My thought was that I could patent the idea and then bid the rights out or license them. I had looked deep into my pockets and found that with the slightly less than $200 that I had on me at the time setting up a genetic research facility was out of the question.

As it turns out, patenting the idea was out of the question. For one thing, the idea was too generic.

Granted, that would not matter if I was a patent-trolling company – their entire business model is based upon generic ideas – but, alas, I was not. Then, there was the cost. Seriously, look up the fees that the government will charge you just to file a patent. You will quickly realize that there's something wrong with the system. I didn't feel like blowing thousands of dollars submitting an idea that they might reject anyway, so I decided that I would write a book about it.

And then I sat on the idea. I went ahead and finished the chapter in which I introduce my idea, and planned to write the book about RSCNT whenever I finished my science fiction book, publishing both simultaneously. Kind of an 'if you found RSCNT interesting, read more about it in the accompanying non-fiction book, *One Cure*' sort of thing. But I ran into a problem. For months now I've been unable to write a single word of fiction. I'm stuck at a little under 100,000 words (which is somewhere around halfway judging from my outline and how long each chapter has turned out to be). I've been *preoccupied* with the cancer idea.

So, at the end of the day, this essay is my attempt to get over my writer's block. Or rather, procrastination. I wouldn't so much call it writer's block since I know exactly what I want to say in the next chapter, but keep putting off typing the words until I have this done and out there for the world to see. So, hopefully, if you like the ideas found within this work, then I will have a work of science fiction for you to read in the next few months to a year. Unless some other wonderful idea decides to become an albatross around my neck.

About the Author

John R. Freeman II was born in 1984 in Paris, Texas. Unlike many millennials (or whatever generation he's in – it changes based upon the source), he has been continuously employed ever since he got his first job when he was 16. He completed a Bachelor's Degree in History with a minor in English from Columbia College in 2013 while employed full time. Why did he wait so long to do that? Because unlike the rest of his generation he only got a degree when it became a prerequisite for advancement at his job. He saw how other people his age were sitting around under/unemployed with expensive and useless degrees and wanted no part of it, choosing to work instead.

He has many hobbies. His favorite is stargazing. He owns an Orion 10" Dobsonian telescope that he tries to use as much as possible. His favorite celestial objects are Jupiter, Saturn, and M13. Reading is probably his second favorite hobby. Others include video games (real ones, not the bullshit you play on your phone), writing (not academically – college cured him of that), and spending hours deep into the rabbit hole known as the internet, where he prefers to focus on libertarianism, the latest scientific research and news, and deepening his knowledge of history.

He is also single... *ladies*. He prefers to believe that his hobbies and interests have nothing to do with that.

You can keep up to date with his writing progress, random thoughts, and future projects at his new and currently very sparse blog: www.johnrfreemanblog.wordpress.com

www.ingramcontent.com/pod-product-compliance
Lightning Source LLC
Chambersburg PA
CBHW070821180526
45168CB00002B/702